ISBN 978-0-265-94259-8
PIBN 10913311

# Historic, archived document

Do not assume content reflects current
scientific knowledge, policies, or practices.

# INDICATORS OF FOREST LAND CLASSES IN
# AIR-PHOTO INTERPRETATION OF
# THE ALASKA INTERIOR

H. J. LUTZ
AND
A. P. CAPORASO

## ALASKA FOREST RESEARCH CENTER

U. S. DEPARTMENT OF AGRICULTURE, FOREST SERVICE

JUNEAU, ALASKA

R. F. TAYLOR, FORESTER IN CHARGE

# TABLE OF CONTENTS

# VEGETATION AND TOPOGRAPHIC SITUATION AS INDICATORS OF FOREST LAND CLASSES IN AIR-PHOTO INTERPRETATION OF THE ALASKA INTERIOR

by

H. J. Lutz [1]

and

A. P. Caporaso [2]

Mature forest stands usually provide a reliable basis for evaluating the site quality or potential productivity of forest land. When such are present, classification of forest land is not a difficult problem. In a region where the natural vegetation has been subject to widespread disturbance, however, classification of forest land into productivity classes, even if broad, becomes a complex process requiring use of all available evidence. Such a region is the Alaska interior. Widely ranging and repeated fires have in the past sixty years destroyed great areas of forest with the result that much of the country now lacks mature timber. The recently burned lands are seldom barren; almost invaribly they are in the process of revegetation, supporting reproduction of tree seedlings or sprouts, a growth of tall or low shrubs, or herbaceous vegetation. Areas swept by forest fires more than 20 or 30 years ago now commonly support trees of sapling and pole-timber size.

During the summer of 1957 the senior author sought criteria, or guides, applicable to air photos, by which the burned areas could be separated into broad forest land classes, based on relative productivity of wood. The objective was to develop a basis for recognizing land capable of producing industrial wood and land not capable of such production. These criteria will be used on the Forest Service's project - The Forest Survey of Alaska. Infrared aerial photography modified by a minus blue filter and flown to a scale of 1:5,000 is used throughout the survey area as well as in this study.

During the course of the work topographic situation and forest vegetation conditions were observed at 123 stations or sites, representing a wide range of land and vegetation classes. At most of the stations soil borings were made to ascertain drainage conditions and depth to frozen ground. Information on the vegetation usually included species composition, and height, diameter, and age of dominants. The sites were classified in the field as commercial forest land or as protection forest land, the latter being regarded as not capable of producing crops of wood having industrial potential.

---

[1]  Alaska Forest Research Center, U. S. Forest Service, and School of Forestry, Yale University.
[2]  Alaska Forest Research Center, U. S. Forest Service.

Protection forest land is so called because the principal role of its vegetation cover is protection of other values. Among these are protection of the land from erosion and thermokarst phenomena; maintenance of hydrologic conditions normal to the region; provision of habitats for wildlife; and protection of the scenic values inherent in natural landscapes.

It should be stated that field classification into the two broad forest land classes was to some extent subjective. In certain areas the vegetation present represented an early stage of successional development following fire and in others the stands were in the early pole stage. Table 1 gives the minimum height and diameter attained at various ages by trees on forest land regarded as capable of producing commercial stands. Minimum basal area values for white spruce and black spruce trees 5 inches d.b.h. and over in stands of commercial quality are also given for various ages. The values given should be viewed only as proximate and not as well-established mensurational specifications. As in all classifications relating to vegetation, one encounters border-line situations. In the majority of instances, however, forest land in the Alaska interior can be assigned to one of the two broad productivity classes with a reasonable degree of assurance.

Table 1.--Minimum height and diameter (all tree species) and basal area (black and white spruce, 5 inches d.b.h. and over) on commercial forest land.

| Age, years | Height feet | D.B.H., inches | Basal area, square feet per acre |
|---|---|---|---|
| 40 | 20 | 3.0 | 25 |
| 80 | 30 | 4.0 | 40 |
| 120 | 35 | 5.0 | 60 |
| 160 | 45 | 6.0 | 75 |

## VEGETATION AS AN INDICATOR OF FOREST LAND CLASSES

Plants have fairly definite requirements for their development; sites which meet the requirements of a given species represent potential habitats whereas habitats not meeting the requirements are unavailable for the species concerned. Thus, certain plants may, by their presence or absence, serve to indicate general or specific site conditions.

During the course of the work reported here primary attention was devoted to tree species because they can usually be identified on air photos. Consideration was also given to the indicator significance of shrub communities (principally willows and alders) and wet meadow communities.

## White spruce, *Picea glauca* (Moench) Voss

Recognition of spruce, as such, on air photos presents no difficulty but problems are encountered in differentiating white spruce and black spruce, *Picea mariana* (Mill.) B.S.P. Raup and Denny (19, p. 120) concluded that "On aerial photographs the two spruces can be distinguished only by their topographic position, not on the basis of form." It is certain that form alone is not a completely reliable criterion for distinguishing white and black spruce on air photos but this is also true even if one is working on the ground. All available criteria should be employed and form is only one of these. On air photos white spruce crowns register as dark grey to black and tend to be rather uniformly spire-shaped, or conical. They cast black, narrowly-triangular shadows (figs. 1 and 2). Mature white spruce trees are generally much taller than mature black spruce; the latter are usually less than 30 to 35 feet. When occurring with black spruce the white spruce trees are the tallest in the stand, often exceeding the height of the black spruce by 50 percent, or more. White spruce trees, excepting those on slumping or caving stream banks and lake shores, tend to be nearly vertical whereas black spruce are often inclined. On recently burned areas where snags are still standing or tree trunks can be observed on the ground (as, for example, in the Matanuska River Valley, at Mile 85 on the Glenn Highway), it may be possible to recognize white spruce as a former occupant of the land (figs. 3 and 4). Useful indications may be supplied by (1) height of standing snags or length of their shadows and length of boles on the ground (white spruce usually more than 40 feet; black spruce usually less than 30 feet), and (2) appearance of the standing snags (white spruce usually with branches, boles straight; black spruce with branches burned off or at least very narrow crowned, boles inclined). Admittedly, the appearance of the standing white and black spruce snags varies with the intensity of the forest fire.

Below the subalpine zone (which is generally above 2,500 feet elevation), white spruce is regarded as a fairly reliable indicator of commercial forest land. Stoeckeler (26, p. 54) noted that "It rarely occurs in waterlogged swamps or dry sandy slopes". Observations by the writers lead to the conclusion that white spruce is more sensitive to poorly drained, cold, wet soils with shallow depth to frozen ground than it is to dry soils. It is usually absent from the poorest black spruce stands on slopes with a northerly exposure and in very poorly drained lowland or upland situations. White spruce may occur as scattered individuals (ten or more

Figure 1.--White spruce stand with dominants 13 to 17
inches d.b.h., height 85 to 100 feet, age 150 years.
Permafrost, if present, more than 42 inches below
the surface. Adjacent to O'Brien Creek, about 12
miles above its junction with the Fortymile River.

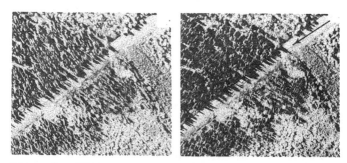

Figure 2.--White spruce with typical spire-shaped or
conical crowns and narrow, triangular shadows. Adjacent to the Chena River east of Fairbanks.
Elev. 500 feet.

Figure 3.--Recent burn (16 years ago) with standing and
fallen snags of white spruce. The larger snags are
12 to 16 inches in diameter; this is commercial forest
land. Matanuska River Valley.

Figure 4.--Recent burn in a white spruce stand. Merch-
antable timber has been salvaged. Dead branches are
still present on standing and down snags. Hardwoods
restocking the area. Note charred appearance of the
ground which is intensified by infrared photography.
Fairbanks. Elev. 500 feet.

trees per acre) in some of the better black spruce stands and its presence may be regarded as an indication of commercial forest land. When white spruce does occur in mixture with black spruce, paper birch, Betula papyrifera Marsh. may also be expected.

In the subalpine zone white spruce, pure or in mixture with black spruce, may form open, very slowly growing stands that locally have been termed "woodland". The trees are short boled, have very rapid taper, and carry their crowns nearly to the ground. Ground cover between the widely spaced trees usually consists of a heavy growth of lichens, mosses, and low shrubs. The most obvious and abundant of the latter are the bog birch, Betula glandulosa Michx.; grayleaf willow, Salix glauca L.; bog bilberry, Vaccinium uliginosum L.; and Labrador-tea, Ledum palustre L. ssp. groenlandicum (Oeder) Hult. On air photos the ground vegetation shows up as grey patches between the scattered dark spruces. These stands are not capable of commercial wood production.

### Black spruce, Picea mariana (Mill.) B. S. P.

Black spruce crowns tend to be cylindrical and, as pointed out by Stoeckeler (27, p. 10), the dark shadows cast tend to be "cigar-shaped". Although recognizing the difficulty of differentiating the spruces on air photos, Raup and Denny (19, p. 123) mentioned that there are . . . . . . . . "slight differences between the tops of the two species, for the black spruce has a common habit of "bunching out' at the very top, so that the point of light reflected is likely to be a little larger than with white spruce." Stoeckeler (27, p. 10) also noted the "knob-like development on the crown tip" and regarded it as common in mature black spruce. Black spruce trees, even when mature, generally do not exceed 30 to 35 feet in height. Black spruce trees often have inclined or crooked boles owing to instability of the substratum in which they are rooted. In open, muskeg areas black spruce trees often occur in distinct clumps as a result of vegetative reproduction by layering. In the open areas, lichens, mosses and low shrubs are clearly visible as light-toned patches on air photos (figs. 5 and 6).

Black spruce is the principal tree in the Alaska interior forming forest stands incapable of commercial wood production. This is not to say that all black spruce stands in Alaska are in this category, but unfortunately most of them are. The species, in practically pure stands, occurs principally on sites that are cold, wet, poorly aerated, and with shallow depth to frozen ground (commonly not more than 12 to 20 inches below the top of the moss and lichen layer). On these sites height growth is very slow (usually less than 35 feet at 120 years age), tree diameters are small (usually less than 5 inches at 120 years), and volumes as indicated by basal area are low (usually less than 60 square feet per acre at 120 years).

Figure 5.--Black spruce muskeg forest on wet, flat land without permafrost. The dominants average 2 to 3 inches in diameter, are 20 to 30 feet in height and 140 years old. Near Houston, west of Wasilla.

Kellogg (13, p. 17) regarded black spruce in Alaska as "characteristically a swamp or muskeg tree". According to Dachnowski-Stokes (4, p. 3) the word muskeg is of Indian (Algonquian) origin and ". . . . . applied in ordinary speech to natural and undisturbed areas covered more or less with sphagnum mosses, tussocky sedges, and an open growth of scrubby timber."

In 1945, "Forest cover types of western North America", (2) a report of the committee on western forest types, was published by the Society of American Foresters. The black spruce type was described as . . . . .

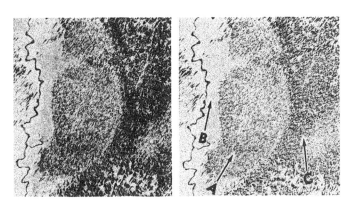

Figure 6.--Black spruce muskeg-A. Note the lighter-toned shrubby vegetation and grasses between trees. Wet meadow-B. Commercial white and black spruce-C. The borders of wet meadows and muskegs follow the contour lines. Near Soldotna on the Kenai Peninsula. Elev. 100 ft.

"commercially unimportant". . . . and . . . . "always occupying cold, wet, acid peat soils commonly called muskegs in the North". . . . That black spruce is regarded as typically a tree of muskegs is further indicated by Taylor and Little (30), Hansen (9) and Drury (5). The present writers would designate as black spruce muskeg all forest lands in the Alaska interior bearing black spruce growing in cold, wet, peat soils and accompanied by characteristic mosses, lichens, and herbaceous and shrubby vegetation. Included in this category is much of the land at relatively high altitudes bearing sparse stands of black spruce. Sometimes these open stands are referred to as "woodland". With only rare exceptions black spruce muskeg forest is incapable of commercial wood production because of the small size of the trees, the excessively slow growth, and the low volume of wood in mature stands.

Occasionally black spruce occurs in mixture with white spruce, Alaska paper birch, quaking aspen, Populus tremuloides Michx., or balsam poplar, P. balsamifera L. When this situation prevails the site is better drained and frozen ground is not as near the surface; commercial forest land is usually indicated.

Paper birch, Betula papyrifera Marsh.

Foliage tone, crown form, and the white boles of the paper birches [1] usually provide a basis for recognition on air photos. Of the three commercial hardwoods, birch registers the lightest on air photos, sometimes almost white. Mature crowns are broad, rounded, and compact. Stand texture of pure, young, even-aged stands appear fine to medium. Overmature stands and older birch mixed with spruce appear coarse in texture. The light tone boles are visible on the photo edges where the trees are viewed more from the side than the top.

The paper birches have long been associated with relatively good forest land in Alaska, Kellogg (13, p. 17), Eakin (6), and Bennett and Rice (1, p. 178). Both Rockie (22, p. 3) and Moffit (15, p. 83) recognized paper birch as indicative of fair to good soil drainage. Pure stands of paper birch, or mixed stands (figs. 7 and 8), in which paper birch represents an important element, indicate commercial forest land. The presence of paper birch (15 to 20, or more, trees per acre) in black spruce stands indicates that the land is probably capable of producing commercial wood. The paper birches are completely or essentially lacking in black spruce muskeg forest.

---

[1] The three varieties of Betula papyrifera in Alaska are not readily recognized and are here treated merely as "paper birch."

Figure 7.--A mixed stand of white spruce, black spruce,
quaking aspen, and paper birch on a 15 percent slope
with a S 70° W exposure. Dominants are up to 12
inches in diameter, about 100 years old. West Fork
of the Fortymile River.

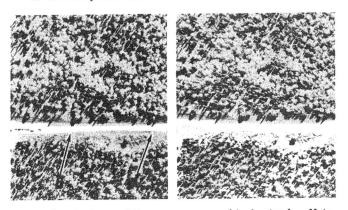

Figure 8.--Mature white spruce-paper birch stand. Note
the excellent contrast in tone and crown shadows between species, spruce-A, and birch-B. Near Soldotna
on the Kenai Peninsula. Elev. 100 ft.

Quaking aspen, <u>Populus</u> <u>tremuloides</u> Michx.

Foliage tone of quaking aspen is usually darker than that of paper birch
as also are the boles of the aspen.  Crowns of mature aspen tend to be
rounded whereas those of balsam poplar tend to be pointed.  For compar-
able heights aspen crowns are smaller in diameter than either birch or
cottonwood.  Aspen commonly, if not usually, occurs in pure, dense, even-
aged stands.  On air photos individual crowns are more difficult to distinguish than either birch or balsam poplar as aspen has less foliage in
the crowns.  Individual crowns often appear fuzzy due to the constant
trembling of the leaves.  Pure aspen stands appear soft and fluffy and
almost never have a coarse or ragged texture.  In many cases the ground
can be seen through the thin canopy (Figs. 9 and 10).  Fire-killed aspen
may often be distinguished from fire-killed birch by the fact that aspen
characteristically produces abundant root suckers that grow rapidly and
form dense, more or less circular patches surrounding the killed trees.

Figure 9.--Pure quaking aspen stand in which the dominants are 3 to 4 inches d.b.h., 40 to 50 feet in
height and 30 years old.  On a slope having an
exposure of S 20° W and a gradient of 15 percent.
Near Ester, Fairbanks district.

Quaking aspen has unusual value as an indicator of commercial forest
land. It is more sensitive to poorly drained, cold soils with shallow
depth to frozen ground than any other forest tree in the Alaska interior.
Kellogg (13, p. 17) associated it with good forest soils and Stoeckeler
(26, p. 65) observed that "Aspen stands occur mainly on unfrozen, well-
drained residual and loess soils and on the upper portion of colluvial
slopes." Stoeckeler also expressed the view that quaking aspen cannot
tolerate waterlogged soils. The opinion seems to be generally unanimous
that the presence of aspen indicates that frozen ground, if present, is
at least four or more feet below the surface (Kellogg and Nygard, (12,
p. 75); Jenness (11), citing A. Leahy; Frost and Mintzer (7); Raup and
Denny (19) Sager (23); Stoeckeler (27); Hopkins, Karlstrom, and others
(10); and Drury (5, p. 53).

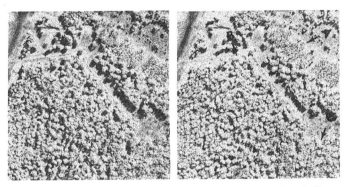

Figure 10.--Pure stand of aspen at A, mature birch-B,
and mixed stand-C. Soldotna. Elev. 100 ft.

The common occurrence of aspen on very steep, south-facing slopes that
are extremely dry and warm marks the species as one of the most xeric
trees in the Alaska interior. Only white spruce approaches aspen in
this respect. On over-steepened slopes (of more than 60 percent) with
a southern exposure aspen does poorly; both height and diameter growth
are excessively slow and the stands will never be sources of commercial wood.

Occurrence of aspen, excepting on overly steepened south-facing slopes,
is a good indication of commercial forest land. The species is more
sensitive than paper birch to poor drainage and shallow depth to frozen
ground. Aspen is almost completely absent from black spruce muskeg
forest. When aspen does occur with black spruce it indicates better
drainage and forest land of commercial potentiality.

## Balsam poplar, _Populus balsamifera_ L.

The tonal appearance of balsam poplar and aspen on airphotos is similar. However, Stoeckeler (27) has suggested that crown form may be helpful in making a separation. In balsam poplar the crown is pointed, whereas in mature aspen it is rounded. Stoeckeler also observed that poplar crowns are denser than aspen crowns. Topographic situation should also prove helpful in separating poplar and aspen stands. The poplar is characteristically a tree of river flood plains, especially on the inner bends of stream meanders, on lake shores, and on sand and gravel flats. Aspen more commonly occurs on upland situations.

Land supporting balsam poplar is frequently under the influence of moving water. Frozen ground, if present at all, is at least three to six feet below the surface. (Sager (23, p. 568); Hopkins, Karlstrom, et al., (10, p. 137). Kellogg and Nygard (12, p. 75) stated that, "Capt. Homer Humphrey, of the Army Permafrost Laboratory, reported that the local inhabitants, when selecting a camp site, have learned to look for a place where aspens grow. He added that under these trees or other species of poplars, the frozen layer was generally 6 feet or more below the surface."

Balsam poplar, pure or in mixture with other tree species, is indicative of commercial forest land (figs. 11 and 12).

Figure 11.--Mature stand of balsam poplar on well-drained river alluvium. Dominants are 20 to 28 inches d.b.h., 95 feet in height, and 170 to 235 years of age. Matanuska River valley, above the mouth of King River.

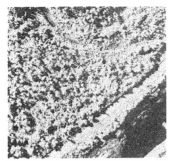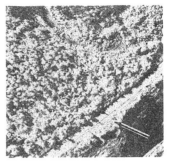

Figure 12.--Sawtimber balsam poplar. Crowns are
pointed as compared to more rounded crowns of
aspen and birch (note crown shadows). Adjacent
to Chena R., Fairbanks. Elev. 500 ft.

## Willows and Alders

There are many species of willows in the Alaska interior but probably
Bebb willow, Salix bebbiana Sarg.; feltleaf willow, S. alaxensis (Anderss.)
Cov.; littletree willow, S. arbusculoides Anderss.; and the Scouler willow,
S. scouleriana Barr., are among the largest and most abundant. Bebb
willow, littletree willow and Scouler willow occur in both lowland and upland situations; feltleaf willow is more commonly found on flood plains
and along streams and on lake shores. The grayleaf willow, Salix glauca
L., is one of the willows most commonly seen in black spruce muskegs.
The most common species of alder are thinleaf alder, A. tenuifolia Nutt.
which commonly forms thickets with the larger willows along streams, fig.
13, Sitka alder, A. sinuata (Reg.) Rydb. and American green alder, A.
crispa (Ait.) Pursh.

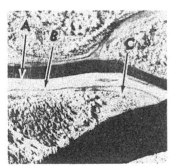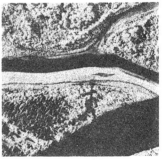

Figure 13.--Tonal and textural differences and loca-
tion serve to distinguish between dense willow, alder,
and balsam poplar on inner banks of stream meanders. A. Light-toned, fine-textured willow. B.
medium-grey, pebbly-grained alder. C. Light-toned,
coarser-grained poplar. Willow and alder on these
sites are temporary and are replaced by poplar and
white spruce. Chena R. near Fairbanks. Elev. 500 ft.

It appears that the presence of tall willows, such as the feltleaf willow, on flood plains, indicates a minimum depth to frozen ground of at least eight feet (Hopkins, Karlstrom, et al., 10). The presence of tall willows (20 or more feet in height) in any situation, whether lowland or upland, indicates forest land with commercial potentiality.

It does not seem possible to assign any general significance to the presence of shrubby willows less than 20 feet in height. One exception may occur in the small stream valleys in mountain areas. Many of these small valley bottoms are occupied by a dense growth of willows six or eight feet in height. There is no evidence that commercial forest stands ever have or ever will occupy these willow-covered valley bottoms. The presence of scattered, dwarfed black spruce either in these willow bottoms, or bordering them, may be taken as further evidence that the land will not support commercial forests.

On upland areas bearing a cover of willows or alders less than 20 feet in height the site must usually be judged by criteria other than the brush cover itself. Criteria that should be helpful may be listed as (1) nature of adjacent forest stands on similar terrain, (2) dead snags and relict trees (individuals or groups) that have escaped fire, (3) topographic situation, and (4) drainage conditions. Some difficulty may be experienced in distinguishing brush from seedling and sapling stands of commercial hardwood species. Tonal differences are slight but generally brush is darker-- medium to dark grey. Multiple brush stems often produce rounded clumps. Foliage is dense with a fine photo texture--never coarse. In general willow clumps are smaller in diameter than alder clumps (figs. 14 and 23).

Figure 14.--Alder clumps on an upland site. Near
Index Lake in the Talkeetna Mts. Elev. 2,500 ft.

## Grass and Sedge Meadows

Within the forested region of interior Alaska one occasionally encounters natural meadows of grasses or grass-like vegetation. As a rule the natural meadows are wet and occur in lowland situations in areas of restricted drainage; commonly they are bordered by black spruce muskeg forest (figs. 6 and 15). These wet meadows are not forest land of any kind. In most cases, there is no evidence that they have supported forest in the past nor that they will in the foreseeable future. Wet lowland meadows can usually be separated from dry meadow areas by the fact that the borders of the former follow rather closely contour lines; the boundaries of dry meadows on uplands often cross contours. Dry meadows are not natural; they generally owe their origin to repeated fires that have exterminated tree growth (fig. 23). Judgment as to the capability of dry meadows for commercial timber production must be based on criteria such as those enumerated above in the closing paragraph on willows.

Figure 15.--Wet meadow on right, black spruce muskeg on
left. The hill and upland in the left background represents commercial forest land supporting white spruce and
paper birch. Here as elsewhere on the Kenai Peninsula
the black spruce type boundaries tend to follow the contours.

Meadows photograph very light on air photos, with no apparent stature, and texture is very fine. Wet meadows register in darker tones than dry meadows on infrared film.

## Type or Stand Boundaries

The nature of the boundaries of vegetation types may sometimes be used in judging site conditions. In most simple terms two categories of stand borders may be recognized: (1) boundaries that follow contour lines and (2) boundaries which do not follow but cross contour lines. Vegetation boundaries that follow contour lines usually indicate that the community or communities involved are under the control of ground water. This is very noticeable in the case of wet-meadow vegetation and, to a lesser extent, black spruce muskeg forest outside the region of permanently frozen ground. The wet meadow-black spruce muskeg boundary is often very clear, closely following the contour. The upper boundary of the black spruce muskeg forest (outside the region of permafrost) likewise tends to follow the contour. Vegetation boundaries that follow contour lines may also be looked for in localities where the soil parent materials were water laid.

Vegetation boundaries that do not follow contour lines but cross them irregularly usually represent edges of old burns. Observation of the oldest adjacent forest vegetation, or remnants of the previous stand that escaped fire, should be helpful in judging site conditions on recent burns. Vegetation boundaries that cross contour lines at approximately right angles fall in a different category. When they occur on slop faces, they are usually associated with erosional features. Strips of aspen, paper birch, or white spruce (pure or in mixture) running up and down north-facing slopes usually mean either (1) shallow gullies or washes with better than average drainage, or (2) slight ridges with better than average drainage (fig. 16).

Figure 16. -- Slight differences in drainage are frequently reflected by the forest composition. Here minor ridges support paper birch on a white spruce slope having a S 80° W exposure. On O'Brien Creek, about 5 miles from its junction with 'he Fortymile River.

-16-

Topographic situation has an important, and frequently controlling, influence on forest productivity in the Alaska interior. This influence arises through the effect of topographic situation on insolation, depth to frozen ground, and drainage. These site factors are all interrelated. Frost and Mintzer (7) remark that "Of all local conditions it is believed that the influence of topographic position is the greatest in determining degree and type of permafrost, either from a field survey or from interpretation of aerial photographs." According to Taber (28, p. 1444), "Forests cover about half the area of perennially frozen ground in Alaska." Some of the relations between topographic situation and forest land classes are shown schematically in fig. 17 and the general distribution of permafrost is indicated in fig. 18.

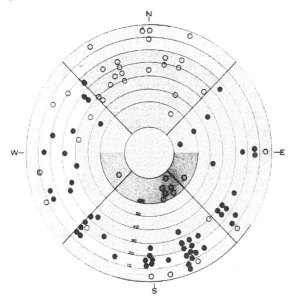

Figure 17.--Relation of forest land classes to slope exposure and slope gradient in the Alaska interior. Slope gradients are shown in percentage classes, less than 10, 11-20, 21-30, etc., and are indicated by concentric zones. Black circles represent commercial forest land; white circles represent land not capable of commercial wood production. Light shading indicates black spruce muskeg forest; dark shading indicates excessively dry land on over-steepened, south-facing slopes. No shading indicates commercial forest land. Essentially flat land, without perceptible slope gradient, is not represented in this scheme.

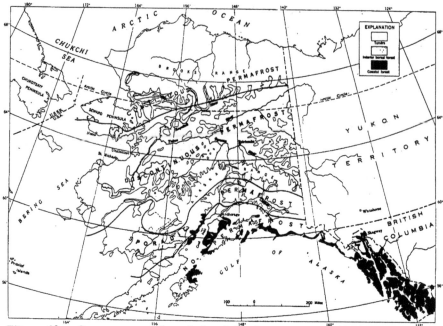

Figure 18.--Occurrence of forested and treeless areas in Alaska,
together with the distribution of permafrost. Treeless areas
are predominantly tundra but also include barren tracts and
grasslands. Courtesy U. S. Geological Survey.

### Exposure or aspect

Slope exposure affects forest site quality, first of all, through its influence on the amount of solar radiation received. Differences of temper-
ature, in turn, affect the depth to perenially frozen ground and this is
related to soil drainage. South-facing slopes receive a maximum of insolation and north-facing slopes a minimum; wet-facing and east-facing
slopes occupy intermediate positions with respect to the amount of heat
received. The effects of exposure on vegetation are greater in the regions of permafrost than in regions to the south. Reed and Harms (21),
working in the Anchorage-Matanuska Valley area concluded that among
the factors controlling the distribution of tree species, exposure played a
minor role except at high elevations. This same situation seems to pre-
vail on the Kenai Peninsula but in the country north of the Coast Range
and north of the Alaska Range, where permafrost is encountered, exposure assumes very great importance.

It is well-known that on south-facing slopes permanently frozen ground is either absent or is encountered at considerable depths below the surface. Bennett and Rice (1, p. 144) observed that during late summer in the Fairbanks section of Alaska, frozen ground was rarely found within 40 inches of the surface on southerly slopes. According to Cressey (3, p. 479), in Siberia "Marked differences are found in the depth of summer thaw between southward-facing slopes and those which receive less of the sun's warmth." Péwé (16) investigated permafrost occurrence in the Dunbar area of Alaska and reported that the contact between the frozen gentle slope area (depositional slope) and the unfrozen steeper slope area was ". . . .generally at a higher altitude on the north-facing slopes than on the south-facing slopes." In the middle Tanana Valley south-facing slopes are generally unfrozen except near the bottoms where gradients are low (Péwé 17, p. 327); Hopkins, Karlstrom, et al. 10, p. 127). Drainage conditions are distinctly better on slopes having a southern exposure than on slopes facing the north; in fact, drainage is excessive on over-steepened south-facing slopes.

Throughout interior Alaska the forest vegetation on slopes with a southern exposure is strikingly different than that found on north-facing slopes (Kellogg, 13, p. 17; Eakin (6); Bennett and Rice 1, p. 178; Rausch, 20, p. 154; Frost, Johnstone, Mintzer, et al., 8, p. 27; and Hopkins, Karlstrom, et al., 10, p. 130). White spruce, paper birch, and aspen, in pure stands or in mixture, are found on south-facing slopes whereas the characteristic cover on north-facing slopes is black spruce muskeg forest (figs. 19 and 20). South-facing slopes having a gradient of more than 10 percent and less than 60 percent represent commercial forest land whereas land having a north exposure is usually incapable of producing commercial wood crops. At slope gradients of more than 60 percent, south-facing slopes are too dry for normal forest growth and are either treeless or support stunted aspen in scattered groups or open stands; these situations cannot be regarded as commercial forest lands. In 1902 Macoun (14) traveled from Whitehorse to Dawson, Yukon Territory, Canada, and observed that slopes along the Yukon River having a southwest exposure and receiving nearly continuous sunshine were covered with grass.

On north-facing slopes frozen ground is usually met at shallow depths. Bennett and Rice (1, p. 178) observed that "Frozen material is reached much nearer the surface on the mossy, northerly slopes, being encountered here usually at depths ranging from about 8 to 24 inches." Péwé (17, p. 327) reported perennially frozen ground extending nearly to the summit of north-facing slopes in the Fairbanks area; Hopkins, Karlstrom and others (10) stated that "In the upland north of the Tanana Valley, silt and bedrock are frozen to unknown depths beneath north slopes." The senior author usually encountered frozen ground at depths of 12 to 20 inches below the top of the moss cover in black spruce muskeg forest on north-facing slopes. As a rule the tree roots are confined to the organic

Figure 19.--The slope on the left (exposure S 45º W, gradient 26 percent) bears a good stand of paper birch.    The
slope on the right (exposure N 25º E, gradient 50 percent) bears a poor black spruce muskeg forest.    The
valley bottom is covered with willows.    Looking up Dead-
wood Creek, tributary to Pedro Creek, in the Fairbanks
district.

soil layer.  If they penetrate mineral soil at all, it is only in the upper-
most few inches.    Taber (28, p. 1463) remarked that "Where slopes are
protected from thawing by a thick accumulation of peat and vegetation, the
slopes become static, and the topography as well as the soil may be described as perennially frozen. "

In the region of discontinuous permafrost the senior author has frequently
encountered small valleys showing pronounced asymmetry.    These valleys
generally had an east-west trend and were occupied by small streams,
without flood plains.    The north-facing slopes were much steeper than the
south-facing slopes.    It appears that this situation is largely the result
of more active physical weathering on the south-facing slope and more
down-slope movement of the soil and rock material.    South-facing slopes
frequently, if not generally, have a deeper mantle of silt; on north-facing
slopes the silt is thinner and in some instances bedrock is encountered a
few inches below the peaty organic layer.    Either erosion must have been

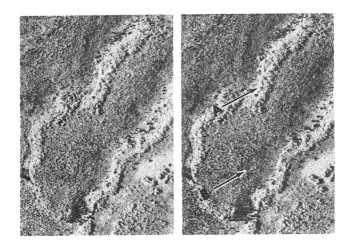

Figure 20.--Site as affected by exposure. A.
Commercial birch on south exposure, and B.
Non-commercial black spruce muskeg on north
exposure. South side of Matanuska R. near
mile 96 on the Glenn Highway. Elev. 1700 ft.

more intense on the north-facing slopes or the mantle of silt has always
been thinner. The writers favor the latter view. Whether one accepts
a dominantly eolian or a dominantly residual origin of the silt, conditions
for accumulation, or formation, appear to have been more favorable on
south-facing than on north-facing slopes.

Drainage on north-facing slopes is usually poor. This results from the
shallow depth to frozen ground and the deep accumulation of peaty organic
matter derived principally from mosses, lichens, and shrubby vegetation.
Black crowberry, Empetrum nigrum L.; Labrador-tea, Ledum palustre L.
ssp. groenlandicum (Oeder) Hult.; bog bilberry, Vaccinium uliginosum L.;
cowberry, V. vitis-idaea L.; and bog birch, Betula glandulosa Michx.,
are among the most common shrubs.

Tree growth is practically restricted to poor, very slowly growing, often
rather open stands of non-commercial black spruce. These north-facing
slopes usually are not capable of commercial timber production.

On slopes facing the east or the west conditions with respect to frozen
ground, drainage, and vegetation stand intermediate between those on slopes
facing the south or north. Presence or absence of permafrost is not as
strongly predetermined on east and west slopes as it is on north and south
slopes. Consequently, sub-stratum conditions affecting drainage may be

-21-

of special importance on moderate slopes with an east or west exposure. For example, it has been observed that such areas underlain with fractured or frost shattered bedrock are unusually well drained and are not frozen to depths normally occupied by tree roots. The slopes facing the east or the west are more nearly like those facing the south than those facing the north. Land having an east or west exposure and a gradient of 10 percent or more, is usually capable of producing commercial crops of wood.

## Degree of slope

Degree of slope is useful in determining broad forest land classes in the Alaska interior. The influence of slope gradient is indirect; it is manifested in soil drainage on one hand and in the position of the permafrost table on the other. The effects are most pronounced when slope gradient is low (less than 10 percent) and when it is high (above 60 percent). Within the slope range from about 10 percent to 60 percent effects on soil drainage and depth to perennially frozen ground are obscure. Likewise indistinguishable are influences on forest vegetation.

The writers have repeatedly observed that when the slope gradient falls below about 10 percent, soil drainage is impeded and frozen ground is often, but not invariably, met at shallow depths of 12 to 20 inches below the top of the moss layer. Péwé (16, 17) recognized a distinction between terrain conditions repreented by erosional slopes and depositional slopes. In discussing boundaries between areas with and areas lacking permafrost he wrote in 1954 as follows: "Generally, however, the contact is near a break in the slope such as the contact between erosional and depositional slopes; where the steeper angle of the hillside gives way to a more gentle slope, drainage becomes sluggish and the water-saturated ground is frozen." The sluggish drainage results partly from the low gradient but a contribution factor is the nature of the sediment forming the depositional slope. This material is usually fine-textured (silt) and possesses a relatively high content of organic matter. On slopes of less than 10 percent, regardless of exposure, the characteristic vegetation is black spruce muskeg forest. The land is not capable of supporting forest growth of commercial potentiality. As a rule, the boundary between the black spruce muskeg forest and the land bearing commercial forest is distinct and easily recognized (fig. 21).

At slope gradients in excess of 60 percent, and with exposures from due west, through south, to due east, drainage is excessive, insolation is high, and frozen ground is either lacking or at such a great depth below the surface as not to affect the vegetation. Land in this category is incapable of commercial forest production; tree growth if present at all, usually consists of very poor aspen. Limiting factors are excessive dryness of the soil, excessive transpiration due to high and nearly continuous insolation, and soil instability due to over-steepening of the slopes.

Figure 21.--Looking along the contour at the contact
between the erosional slope (on the left, with paper
birch) and the depositional slope (on the right, with
black spruce muskeg). Slope exposure S 25° E.
Gradient of erosional slope, 20-30 percent; gradient
of depositional slope, 7 percent. Note fire-charred
spruce snag, 9 inches in diameter, above the birches.
Chatanika River valley.

These over-steepened slopes usually occur along actively eroding streams
where bank cutting is, or has been, in progress. (figs. 22 and 23).


## Valley Bottoms and Lowland Situations

In valley bottoms and other lowland situations the dominant factors in
determining forest land potentiality are soil drainage conditions and depth
to perennially frozen ground. These factors, as stated previously, are
closely interrelated. Conditions that favor good drainage in bottomland
situations include coarse-textured soils and substrata, and proximity to
well-extablished drainage ways. These conditions permit active circulation of water, thawing frozen gound if present and retarding or prevent-
ing formation of new permafrost.

Figure 22.--Over-steepened south-facing slope on the right bearing scattered white spruce and quaking aspen. On the left side of the stream is a black spruce muskeg. Valley of the Fortymile River.

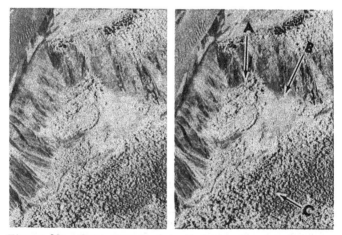

Figure 23.--Over-steepened south-facing slope. A. Noncommercial aspen. B. Dry meadow crossing contour lines. C. Willow clumps. North-side of Matanuska R. at Mile 96. Elev. 1,500 ft.

In the Alaska interior relatively fine-textured soils, occupying situations where gradients are low, usually exhibit poor drainage. Total pore space in these materials is high but the volume of noncapillary pores is low. The result is that movement of water is sluggish. Tedrow and Hill (31) investigated certain well-drained soils on the Arctic Slope in Alaska that they suggest be designated as Arctic brown soil. Although the Arctic brown soil occurs north of the limit of forests, the observations of Tedrow and Hill on the relation between soil texture and permafrost would also seem to apply within the forest area. They state: "Exclusive of frostboil areas, no significant difference in depth to permafrost is noted among silt loams, sandy loams, and loamy sands, but where a large percentage of coarse material is present, such as gravelly outwash, the depth to permafrost tends to increase markedly."

Coarse-textured soils and substrata may be expected in areas of outwash deposited by large streams, and in gravel terraces bordering streams. Alluvium along small streams of low gradient is likely to be fine textured.

It has long been recognized that in northern regions forest growth is usually best along streams and drainage ways. Schrenk (25, p. 254) traveled up the Kolva River (tributary to the Usa River) in Russia and noted that the forest was confined to a belt along the stream; as he went north the width of the forest strip narrowed. At its outer edges the forest gave way to tundra. Similar observations in Russia were made by Pohle (18) on the Kanin Peninsula, by Tanfil'ev (29) in the Timan tundra region, and by Sambuk and Dedoff (24) in the Pechora (Petschora) tundra region. Along streams the vigorous circulation of water through relatively permeable sands and gravels usually accounts for the thawed ground and the favorable conditions for forest growth. Some of the best commercial forest land in the Alaska interior occurs along stream courses (fig. 24). Thawed ground is also found adjacent to recently abandoned river channels,

Figure 24.--White spruce, paper birch and quaking aspen in the foreground; pure white spruce stands on terraces in the background. Well-drained alluvium along the larger rivers represents excellant forest land. Looking easterly across the Copper River Valley, about two miles upstream from Copper Center.

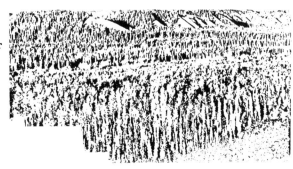

sloughs, and lakes. The influence of drainage and circulation of water on permafrost is generally recognized and has been discussed by various writers on Alaskan conditions, for example Taber (28), Péwé (17), and Hopkins, Karlstrom, and others (10).

Flat valley floors and other lowland situations without well-developed drainage ways usually represent non-commercial forest land. If such sites bear tree growth it is usually a black spruce muskeg forest; occasionally they are treeless.

It should be recognized that even in regions outside the areas of permafrost, soil drainage is tremendously important in determining whether land can or cannot produce crops of commercial wood. Black spruce muskeg forest occurs on poorly drained bottomland areas without permafrost as well as in those situations where frozen ground is within 12 to 20 inches of the surface.

## SUMMARY

Widespread and repeated forest fires have, in the past sixty years, destroyed great areas of forest in the Alaska interior. Classification of these burned lands into broad productivity classes--commercial forest land and land not capable of commercial wood production--involves use of all available criteria. Vegetation and topographic situation may be used by air photo interpreters as indicators or guides in recognizing these forest land classes.

White spruce is a fairly reliable indicator of commercial forest land. It is relatively sensitive to poorly drained, cold, wet soils and is usually absent from black spruce muskeg forests. Under white spruce the minimum depth to permafrost (if present at all) is generally 24-36 inches.

Black spruce is a characteristic muskeg tree in Alaska, usually occupying cold, wet, poorly drained soils. Permafrost is commonly only 12 to 20 inches below the top of the moss and lichen layer. Black spruce muskeg forest on north-facing slopes, at relatively high altitudes, and on land with a slope of less than 10 percent is not capable of commercial wood production.

The paper birches are associated with fair to good drainage conditions and commercial forest land. Under paper birch the minimum depth to permafrost (if it is present at all) is usually 36 to 48 inches.

Quaking aspen is a particularly useful indicator species. It is more sensitive to poorly drained, cold soils with permafrost at shallow depth than any other tree species in Alaska. The presence of aspen indicates that frozen ground, if present, is at least 48 inches below the surface. Quaking aspen is an excellent indicator of commercial forest land. The only exception is when the species occurs on over-steepened south-facing slopes of 60 percent, or more, gradient.

Balsam poplar, characteristically a species of river flood plains, is an indicator of commercial forest land. Where balsam poplar occurs permafrost is either absent or at least three to six feet below the surface.

Tall willows (20 or more feet in height) are believed to indicate land having the potentiality of commercial wood production.

Wet grass and sedge meadows occupy non-forest land.

The nature of the boundaries of vegetation types may be used in judging site conditions. Boundaries that follow contour lines usually indicate that the communities involved are under the control of ground water. Vegetation boundaries that do not follow contour lines but cross them irregularly usually represent edges of old burns. Type boundaries on slope faces that cross contour lines at approximately right angles are usually associated with erosional features. They may mean either shallow gullies or washes with better than average drainage or slight ridges with better than average drainage.

Slope exposure exerts a powerful influence on soil conditions, expecially drainage and depth to permafrost, and forest vegetation in the Alaska interior. South-facing slopes are usually most favorable for forest growth; north-facing slopes are usually least favorable. Slopes with an east or west exposure exhibit conditions intermediate between those found on south-facing and north-facing slopes. Within the permafrost region, north-facing slopes usually support black spruce muskeg forest and are incapable of producing crops of commercial wood. Over-steepened slopes (60 percent gradient, or more) having a southern exposure are too dry to support commercial forest stands. They are often treeless or bear a scattered growth of poor aspen. Land below the subalpine zone having a slope exposure of east, south, or west and a slope gradient of not less than 10 percent nor more than 60 percent may be regarded as commercial forest land.

Degree of slope is particularly important at low gradients (less than 10 percent) and at high gradients (more than 60 percent). When slope gradient falls below about 10 percent, soil drainage is impeded and frozen ground is commonly encountered at shallow depth (10 to 20 inches). Black spruce muskeg forest usually occupies such land. At high slope percentages (over 60), land having a southern exposure is too dry for normal forest growth.

In valley bottoms and other lowland situations soil drainage and depth to permafrost are the principal factors influencing forest growth. Situations where soils and substrata are coarse-textured and adjacent to well-established drainage ways represent commercial forest land. These conditions are found most often along large streams with flood plains. Flat valley floors and other lowland situations without well-developed drainage ways usually represent non-commercial forest land.

## References cited

(1) Bennett, Hugh H. and Thomas D. Rice
     1919. Soil reconnaissance in Alaska, with an estimate of the
           agricultural possibilities. U. S. Dept. Agr., Bur.
           of Soils. Field operations of the Bur. of Soils, 1914,
           16th report, by Milton Whitney. pp. 43-236.

(2) Committee on western forest types
     1945. Forest cover types of western North America. 35 pp.
           Soc. Amer. For., Washington, D. C.

(3) Cressey, George B.
     1939. Frozen ground in Siberia. Jour. of Geol. 47:472-488.

(4) Dachnowski-Stokes, A. P.
     1941. Peat resources in Alaska. U. S. Dept. Agr., Tech.
           Bul. 769, 84 pp.

(5) Drury, William H., Jr.
     1956. Bog flats and physiographic processes in the upper Kus-
           kokwim River region, Alaska. Contributions from
           the Gray Herbarium of Harvard University. No. 178,
           130 pp.

(6) Eakin, Henry M.
     1913. A geologic reconnaissance of a part of the Rampart quad-
           rangle, Alaska. U. S. Geol. Surv. Bul. 535, 38 pp.

(7) Frost, Robert E., and Olin W. Mintzer
     1950. Influence of topographic position in airphoto identification
           of permafrost. Highway Res. Board, Bul. 28, pp.
           100-121.

(8) Frost, Robert E., James G. Johnstone, Olin W. Mintzer, Merle Par-
           vis, P. Montano, Robert D. Miles and James R. Shepard.
     1953. A manual on the airphoto interpretation of soils and rocks
           for engineering purposes. Purdue Univ. School of Civil
           Engin. and Engin. Mech. ix + 206 pp.

(9) Hansen, Henry P.
     1953. Postglacial forests in the Yukon Territory and Alaska.
           Am. Jour. of Science 251:505-542.

(10) Hopkins, David M., Thor N. V. Karlstrom, and others.
     1955. Permafrost and ground water in Alaska. U. S. Geol.
           Surv. Prof. Paper 264-E. iii + 113-146 pp.

(11) Jenness, John L.
    1949. Permafrost in Canada. Origin and distribution of per-
        manently frozen ground, with special reference to
        Canada. Arctic 2(1):13-27.

(12) Kellogg, Charles E. and I. J. Nygard
    1949. Soils of Alaska. In: Report on exploratory investiga-
        tions of agricultural problems of Alaska. U. S. Dept.
        Agr. Misc. Pub. 700, pp. 25-86.

(13) Kellogg, R. S.
    1910. The forests of Alaska. U. S. Dept. Agr., For. Service,
        Bul. 81, 24 pp.

(14) Macoun, John
    1903. The climate and flora of the Yukon District. In: Summary
        report of the operations of the Geological Survey of
        Canada for the cal. year 1902. Geol. Survey of Canada.
        Ann. report (New series) Vol. 15, 1902-3, pp. 38A-53A.
        1906.

(15) Moffit, Fred H.
    1954. Geology of the eastern part of the Alaska Range and adjacent
        area. U. S. Geol. Surv. Bul. 989-D, v + 63-218 pp.

(16) Péwé, Troy L.
    1949. Preliminary report on permafrost investigations in the Dun-
        bar area, Alaska. U. S. Geol. Surv. Circ. 42, 3 pp.

(17) Péwé, Troy L.
    1954. Effect of permafrost on cultivated fields, Fairbanks area,
        Alaska. U. S. Geol. Surv. Bul. 989-F, iv + 315-351 pp.

(18) Pohle, Richard
    1901. Pflanzengeographische Studien über die Halbinsel Kanin und
        das angrenzende Waldgebiet. Inaugural--Dissertation
        zer Erlangung der Doctorwürde. . . .der Landes--
        Universität Rostock. 112 pp.

(19) Raup, High M. and Charles S. Denny.
    1950. Photo interpretation of the terrain along the southern part
        of the Alaska Highway. U. S. Geol. Surv. Bul. 963-D,
        iv + 95-135 pp.

(20) Rausch, Robert
    1951. Notes on the Nunamiut Eskimo and mammals of the Anaktu-
        vuk Pass region, Brooks Range, Alaska. Arctic 4(3):
        147-195.

(21) Reed, John C., Jr. and John C. Harms
  1956. Rates of tree growth and forest succession in the Anchor-
      age-Matanuska Valley area, Alaska. Arctic 9(4):238-
      248.

(22) Rockie, W. A.
  1946. Physical land conditions in the Matanuska Valley, Alaska.
      U. S. Dept. Agr. Soil Conserv. Serv. Physical Land
      Surv. No. 41, 32 pp.

(23) Sager, R. C.
  1951. Aerial analysis of permanently frozen ground. Photogramm.
      Engin. 17:551-571.

(24) Sambuk, F. A. and A. W. Dedoff
  1934. Die Unterzonen der Petschora-Tundren. /Russian text, pp.
      29-50/. Acta Instituti Botanici Academiae Scientiarum,
      Series III, 1933. Geobotanica, Fasc. 1.

(25) Schrenk, Alexander Gustav.
  1949. Reise nach dem Nortosten des europaischen Russlands,
      durch die Tundren der Samojeden, zum Arktischen Ural-
      gebirge, auf Allerhöchsten Befeho für den Kaiserlichen
      botanischen Garten zu St. Petersburg im Jahre 1837
      ausgeführt. Erster Theil. Historischer Bericht. xliv
      + 730 pp. Heinrich Laakmann, Dorpat.

(26) Stoeckeler, E. G.
  1949. Identification and evaluation of Alaskan vegetation from air
      photos with reference to soil, moisture, and perma-
      frost conditions. A preliminary paper, Dept. Army,
      Corps of Engin., St. Paul District, 103 pp.

(27)
      ————————
      1952. Trees of interior Alaska, their significance as soil and
          permafrost indicators. Corps of Engin. U. S. Army,
          Investigation of military construction in arctic and sub-
          arctic regions. Prepared by St. Paul District Corps of
          Engin. for Office of the Chief of Engin., Airfields Branch,
          Engin. Div., Mil. Construction, 25 pp.

(28) Taber, Stephen
  1943. Perennially frozen ground in Alaska: its origin and history.
      Bull. of the Geological Soc. of Am. 54:1433-1548.

(29) Tanfil'ev, G. I.
  1911. Die polare Grenze des Waldes in Russland, nach Untersuch-
      ungen in der Tundra der Timan--Ssamojeden. /Text in
      Russian; German summary, pp. 277-286/ 286 pp. Odessa.

(30) Taylor, Raymond F. and Elbert L. Little, Jr.
      1950.   Pocket guide to Alaska trees.   U. S. Dept. Agr., Agr.
              Handbk 5, 63 pp.

(31) Tedrow, J.C.F. and D. E. Hill
      1955.   Arctic brown soil.   Soil Science 80:265-275.

GPO 991400

Lightning Source UK Ltd.
Milton Keynes UK
UKHW020217030119
334668UK00005B/171/P